THE DUTCH POLDER MODEL IN SCIENCE AND RESEARCH

Royal Netherlands Academy of Arts and Sciences
PO Box 19121, NL-1000 GC Amsterdam
T +31 (0)20 551 0700
knaw@knaw.nl
www.knaw.nl

PDF available at www.knaw.nl

Editing: Peter Vermij
Research: Robbin te Velde
Info graphics: Titia Lelie
Layout and design: Ellen Bouma
Illustrations: Geert Gratama
Translation: Stuart Doughty

ISBN 978-94-629-881-63

This publication is printed on paper that complies with the ISO
9706:1994 standard for document storage and preservation.

Original title: *Wetenschap in Nederland: Waar een klein land
groot in is en moet blijven*. Preferred citation: Van Dijck and Van
Saarloos (2017). *The Dutch Polder Model in science and research*,
Amsterdam.

KONINKLIJKE NEDERLANDSE
AKADEMIE VAN WETENSCHAPPEN

The Dutch Polder Model in science and research

What allowed the Netherlands to punch above its weight?
How should the country build on that achievement?

José van Dijck and Wim van Saarloos

Preface

Scientific research in the Netherlands has an excellent track record. Dutch researchers, universities and knowledge institutes reside at or near the top of global and European rankings. The country's science system has managed to translate uniquely Dutch characteristics, including its 'Polder model', into a success story it can be proud of.

But today's rankings and citation impact scores are the result of hard work and substantial investments in the past. In a way, we are looking in the rear-view mirror. Ahead the road looks a bit bumpier.

Almost every day we talk to scientists, both young and at more senior levels, who experience the growing strain on the Dutch research system. Cracks are beginning to appear, due in part to outside forces, but also partly because of how research communities have responded to those forces.

Many of us feel the time has come to turn the tide. With the economic upturn bringing more breathing space, and scientific and societal challenges demanding new solutions, it is imperative that we now make the right choices. The country will continue to have tremendous opportunities—provided it recognises where its unique strengths lie and it is willing to invest in them again.

We would like to thank everyone who shared with us their insights and concerns about the current and future state of science in the Netherlands. This essay is partly the result of intensive discussions within the Board of the Academy. We hope it will contribute to a lively national debate.

A final note: this essay was originally written in Dutch. For the sake of brevity, we have translated the Dutch word *wetenschap* (*Wissenschaften* in German) as 'science'. Please note that wherever we mention 'science', 'science system' or 'scientists', those terms are meant to include the humanities, the social sciences and the medical sciences as well as the natural sciences.

José van Dijck
Wim van Saarloos

28 August 2017

Summary

For many years now, Dutch researchers and the Dutch science system have been performing outstandingly.

The evidence for that excellence lies in indicators such as the number of articles published and the scientific impact of those publications. It is also apparent from the position of our universities in international rankings. And it is shown by the success of Dutch researchers in the international competition for research funding. Whichever benchmark is chosen, the Netherlands ranks remarkably high today. Holland is a small country that is able to compete with larger nations. That is very important for its future economic position and its capacity to help society meet the major challenges it faces.

Today's accomplishments are the fruits of past investments. They are also due to a combination of factors that is unique to the country. The Netherlands is a compact country, situated in a flat delta and surrounded by major European powers. Throughout history it has invested in connections of every kind. Progress has also been helped by what we call 'polder culture'.

An emphasis on equal opportunities and an aversion to strong hierarchy fostered the development of Dutch research across all disciplines. With their desire for consultation, consensus and collaboration, Dutch scientists have shown themselves to be excellent self-organisers. And for centuries they have been eager to look beyond national borders.

The unique science system is also super-efficient: the engine of Dutch science is running in high gear. But a quick glance under the hood shows that the system is in need of a major service.

Investments in research are stagnating in the Netherlands, while countries around us redouble their efforts. The growing numbers of graduate and post-graduate students are difficult to sustain at the present investment level. The budget for fundamental research, a crucial basis, is partially reallocated to research aimed at meeting societal challenges. As a result, researchers can feel pressured to avoid scientific risk, to focus on short-term outcomes, or to be led primarily by quantitative incentives. Collaboration is making way for competition, while a growing share of research budgets ends up with a shrinking number of scientists. The most talented Dutch researchers are lured by attractive opportunities abroad.

Slowly but surely, the 'high plateau' of the Dutch scientific landscape is eroding.

Fortunately, that trend can be reversed. The current economic recovery enables the Netherlands to invest once again in precisely those traits that made Dutch research strong to begin with.

In the years ahead, the Dutch academic community could regain the prospect of a steadily growing budget from both public and private sources, which would allow it to catch up with its neighbours. The community could build on the concept of a virtual 'University of the Netherlands', creating new and nourishing existing connections. The typically Dutch balance between collaboration and competition, between building a broad collective base and rewarding individual talent, could be restored. Talent could be given more breathing space. Researchers could organise themselves to find answers to new questions with multidisciplinary and cross-institutional research—especially in light of the present demand for answers to societal challenges and practical applications of knowledge.

The science system in the Netherlands continues to contain excellent opportunities to promote welfare and prosperity. Dutch research should help formulate responses to major issues facing our society today. It can do so if, as in the past, everyone concerned recognises and endeavours to enhance the unique strengths of the scientific polder landscape, provides its research with sufficient financial resources, and respects the very characteristics and culture that turned Dutch research into a success.

Contents

5. Nurture Dutch strengths

9

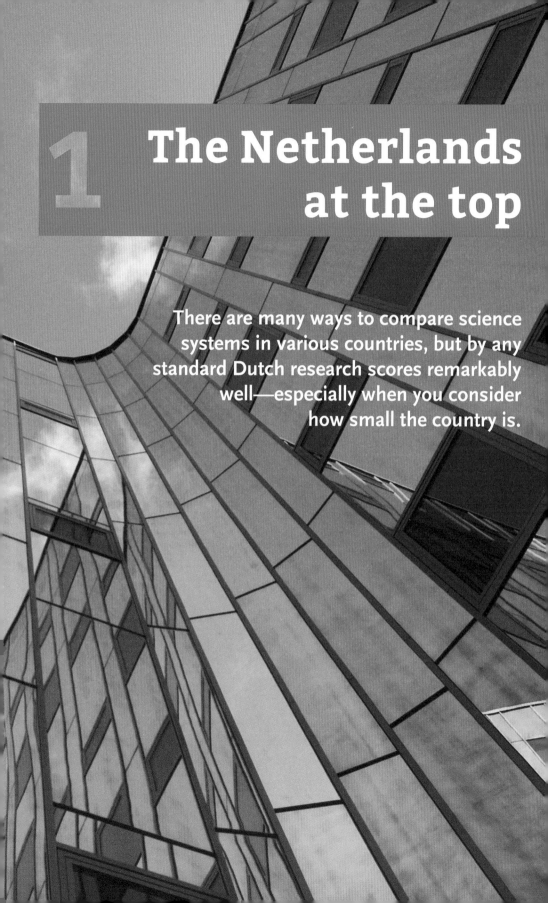

1 The Netherlands at the top

There are many ways to compare science systems in various countries, but by any standard Dutch research scores remarkably well—especially when you consider how small the country is.

A major research producer

DK NL UK DE FR OECD
                        average

SUBSTANTIAL OUTPUT OF SCIENTIFIC ARTICLES

The Netherlands is a major producer of peer-reviewed scientific articles, particularly when the output figures are adjusted for population size. In the period
1996-2015, the Netherlands published twice as many articles per capita as the
average OECD country and significantly more than countries with large scientific
communities such as Germany and France.

Source data: SCImago Journal & Country Rank (www.scimagojr.com) on the basis of Elsevier's
Scopus database; UN World Population Prospects: The 2015 Revision.

The Netherlands is a small country—it is the world's 134th-largest nation in terms of surface area. The country has the world's 65th-largest population. And the economy, measured in terms of 'purchasing power parity', ranks number 27.[1]

But even a small country can be very big in some respects. The Port of Rotterdam is the eighth largest in the world measured by tonnes of freight,[2] and is one of the reasons why the Netherlands is the world's second-largest exporter of agricultural produce.[3]

Few Dutch people realise that this small country is also a major power in the world of science.

One way of measuring its impact is by looking at the number of scientific articles published in international peer-reviewed journals by researchers working at Dutch universities, research institutes and companies.

Naturally, the country cannot measure up to far larger countries with hundreds of millions of inhabitants, like the United States and China. Still, the Netherlands does occupy the fifteenth position in a table ranking countries according to the absolute number of publications.[4]

However, when scientific output is adjusted for the size of countries' populations, the Netherlands leaps to the top of the table, with only Switzerland and the Scandinavian countries publishing as much or slightly more per capita. (In the interests of clarity, the figures in this essay are confined to those of the four countries—Denmark, France, Germany, and the United Kingdom—discussed in chapter 3.)

Compared with the average of all OECD countries, the Netherlands publishes more than twice as many peer-reviewed academic articles per capita. When it comes to research, the Netherlands is a remarkably productive country.

Research with extraordinary impact

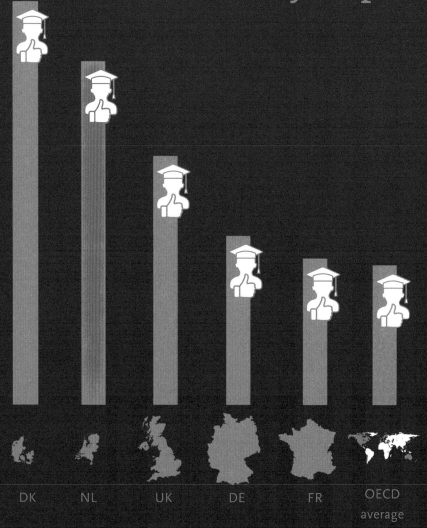

| DK | NL | UK | DE | FR | OECD average |

FREQUENT SCIENTIFIC CITATIONS

Dutch research has a major impact on global science: the citation rate of the country's research is high, particularly when the figures are adjusted for population size. By that measure, between 1996 and 2015, the number of citations of all Dutch research was substantially larger than that of the average OECD country or countries with large scientific communities such as the United Kingdom, Germany or France.

Source data: SCImago Journal & Country Rank (www.scimagojr.com) on the basis of Elsevier's Scopus database; UN World Population Prospects: The 2015 Revision.

Quantity isn't everything, of course. Equally important is whether the work of Dutch researchers is not only regarded as valuable by the peer reviewers for journals, but also by other scientists in their respective disciplines.

An imperfect, but nevertheless frequently used benchmark in this respect is whether published papers are subsequently cited by other researchers.

Citations say something about the quality and novelty of the scientific research and about the scientific impact of that work.

In a table of absolute numbers of citations, the Netherlands occupies a remarkable tenth position worldwide.[5] However, when we adjust the figures for the size of countries' populations, the Netherlands is only surpassed by Switzerland, Denmark and Sweden among the larger research countries— perhaps not coincidentally countries that also invest more money in research and development.

In the period 1996-2015, all Dutch research together garnered twice as many citations per capita as did German and French research. Those last two countries performed slightly better on average than the OECD countries as a whole.[6]

Another way of gauging the citation impact of Dutch research is to measure how often each research paper gets cited on average. The Netherlands ranks second in the world on this benchmark, just behind Switzerland and slightly ahead of Denmark.

Whichever measurement is used, science in the Netherlands is obviously highly regarded elsewhere in the world.

European tokens of appreciation

NL	UK	DK	DE	FR	ERA average
39	25	25	15	14	13

ERC GRANTS PER CAPITA

Between 2007 and 2016 the Netherlands secured 39 prestigious European Research Council grants per million inhabitants. That was three times as many as the average country in the European Research Area (ERA), but also significantly more than countries with large scientific communities like the United Kingdom, Germany and France.

Source data: ERC, ERC Funded Projects (https://erc.Europe.eu/projects-figures/erc-funded-projects, spring 2017); UN World Population Prospects: The 2015 Revision.

Since 2007, one way to compare the quality of research in the Netherlands with that in other European countries has been to use the judgment of the European Research Council (ERC).

The ERC's goal is to promote the best European research regardless of academic discipline. Researchers from 33 countries can apply for grants on the basis of research proposals. The proposals are assessed by juries composed of leading European scientists using a system of peer review in which the sole criteria is scientific quality. The system was inspired by the system of evaluation employed by the Netherlands Organisation for Scientific Research (NWO).

The awarding of an ERC grant is now recognised worldwide as an important indicator of research quality.

Dutch researchers have performed very well in the strict selection procedure during the 10 years of the ERC's existence. When, as with publications and citations, the number of grants awarded is corrected for population size, the Netherlands stands head and shoulders above most other countries (including large and prestigious countries like Germany, France, the United Kingdom and Denmark) in terms of the number of grants awarded—almost forty grants per million inhabitants.

Of all the countries in the European Research Area only Switzerland and Israel have a better record.

The Netherlands' track record not only applies to the total number of ERC grants; it also extends to each of the five sub-categories: Starting Grants, Consolidator Grants, Advanced Grants, Proof of Concept Grants and Synergy Grants.

"This says a lot about the quality of science in the Netherlands," Robert-Jan Smits, the Director General of Research and Innovation at the European Commission, once said. *"After all, [the ERC juries] select solely on the basis of excellence and the competition is intense."*[7]

17

Excellence across the board

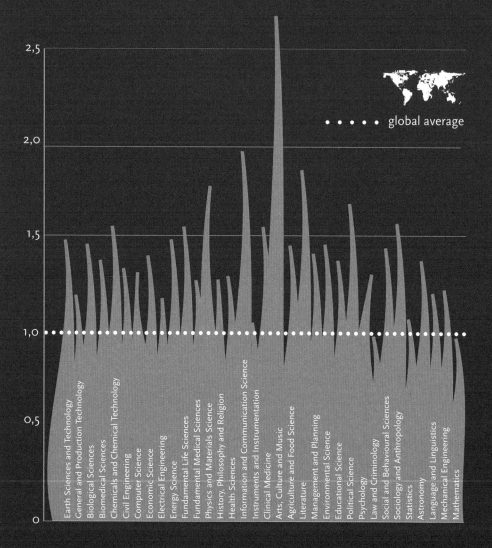

global average

Earth Sciences and Technology
General and Production Technology
Biological Sciences
Biomedical Sciences
Chemicals and Chemical Technology
Civil Engineering
Computer Science
Economic Science
Electrical Engineering
Energy Science
Fundamental Life Sciences
Fundamental Medical Sciences
Physics and Materials Science
History, Philosophy and Religion
Health Sciences
Information and Communication Science
Instruments and Instrumentation
Clinical Medicine
Arts, Culture and Music
Agriculture and Food Science
Literature
Management and Planning
Environmental Science
Educational Science
Political Science
Psychology
Law and Criminology
Social and Behavioural Sciences
Sociology and Anthropology
Statistics
Astronomy
Language and Linguistics
Mechanical Engineering
Mathematics

DUTCH RESEARCH IS UP TO STANDARD IN ANY DISCIPLINE

Citation impacts of in essence all research disciplines in the Netherlands (shown here for 2009-2012 by the top of each 'blade of grass') are above the global 'average' (the dotted line). Dutch research is world-class across the board; there are no 'bald patches' in the lawn. That is one of the strengths of Dutch research.

Source: Science, Technology and Innovation Indicators 2, Centre for Science and Technology Studies (Leiden), Dialogics.

Some countries stand out in particular disciplines but are less prominent in others. That does not hold true for the Netherlands. On the contrary: the country performs well in every discipline. The Dutch scientific landscape has been compared to a 'high plateau' with occasional peaks.

The figure on the opposite page illustrates how the broad range of disciplines in Dutch academia does not have weak spots. When research in the Netherlands is divided into 34 large and small disciplines, the citation impact of each of them can be calculated. The diagram shows the values for the period 2009-2012, with the dotted line representing the global average in each discipline.[8]

It is clear from the figure that in essence there are no below-average disciplines in the Netherlands and that most disciplines perform well above the global average. Positive outliers are arts, culture and music; literature; information and communication sciences; and physics and materials science.

The lawn would appear even flatter if the disciplines are clustered into the four major domains: exact & natural sciences; social sciences and humanities; applied & technical sciences; and healthcare research & medical sciences. In all four domains the citation impact of Dutch research is between 30% and 50% higher than the global average.

The composition of the Dutch academic community does differ in some respects from the situation elsewhere in the world. The Netherlands is relatively active in disciplines such as clinical medicine, health sciences and social sciences, partly because of the relatively large numbers of students in those fields. The size of the exact and technical disciplines are relatively small compared to other countries, which makes their above-average impact even more remarkable.[9]

A highly efficient science system

OECD   FR   DE   DK   UK   NL

Citations/mln. € GERD
2009-2015

0595231½

Citations NL
2009-2015

CITATIONS PER EURO INVESTED

The relatively modest public and private investments in research and development in the Netherlands generate a remarkably high return: Dutch research yielded 74 citations for every million euros invested in the period from 2009 to 2015. No other country in the world got as much citation impact for its investment during that period.

Source: SCImago Journal & Country Rank (www.scimagojr.com) on the basis of Elsevier's Scopus database; Gross domestic expenditure on research and development 2016 according to the OECD's Main Science and Technology Indicators.

In other words, for a small country the Netherlands performs well in science across the board. And in the preceding pages, we did not even taken into account that the Netherlands spends less money on science than many neighbouring countries.

The figure on the opposite page shows clearly that in recent years the Netherlands received excellent value for the money it has invested. For every euro spent on research and development by the government and the private sector, the citation impact of research in the Netherlands was greater than that of any other country in the world. On balance, money invested in research in the Netherlands generated three times as many citations as the same sum invested in the United States. In other words, the engine of Dutch science is running in high gear.

Various scholars and policymakers have commented on that remarkable efficiency. In 2015 Martin Stratmann, the president of the Max Planck Gesellschaft in Germany, said: *"Der Vergleich mit unseren niederländischen Nachbarn zeigt: Wir haben in Deutschland nicht nur ein Problem der finanziellen Grundausstattung unserer Universitäten, wir haben auch ein Effizienzproblem, das sich nur über eine strukturelle Weiterentwicklung der Universitäten lösen lässt."*[10]

In 2013 Nick Fowler, now the Chief Academic Officer at Elsevier, one of the world's largest academic publishers, presented an analysis of the Dutch science system. *"As a relatively small country, the Netherlands is punching well above its weight,"* he found. *"The country is definitely doing something right. The question is: what is it doing right?"* [11]

2   Unique polder foundations

How could a small country like the Netherlands become such a major research power?

Its prime location and systematic investments in connections contributed to its achievements. But an extraordinary culture of openness, collaboration, organisation and aversion to strong hierarchy also contributed to the success.

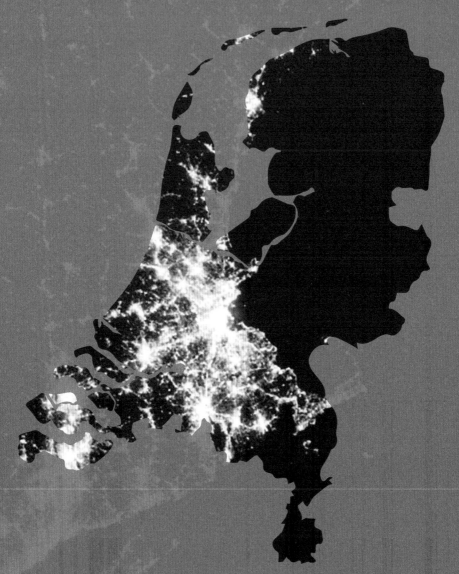

A compact, strategically situated cluster

THE NETHERLANDS AS A METROPOLITAN REGION

A satellite image of the Northeastern United States, with a map of the Netherlands on the same scale superimposed on it. The extended urban agglomeration around Boston (Massachusetts), a city that is home to a lot of scientific research, would cover a large part of the Netherlands.

Source data: Suomi National Polar-orbiting Partnership (Suomi NPP), NASA, 2012.

Numerous factors have helped the Dutch academic community to perform well. One factor is geographical: the Netherlands is a compact, densely populated country, strategically situated between major Western European powers. Such geographic advantages continue to bring opportunities that other countries do not necessarily enjoy.

Seen from space the Netherlands is not much bigger than what we would call a metropolis in other regions of the world. Urban regions in the United States such as New York, Los Angeles and Boston would cover large parts of the country. The Dutch 'Randstad' as a whole is similar in size to large metropolitan regions such as London, Paris or Milan.

But within that small surface area the Netherlands accommodates 13 highly regarded universities,[12] as well as dozens of respected institutes for fundamental research (under the auspices of NWO[13] and KNAW[14]) and applied research (the TO2 institutes,[15] government research institutions[16] and universities of applied sciences (*hogescholen*)). Besides public research there are also many centres of private research—the R&D departments of multinational corporations, innovative medium-sized enterprises and young spin-offs from the universities.

Silicon Valley, the region near San Francisco that has become synonymous with a successful innovative region, would stretch from Nijmegen to Amsterdam on a map of the Netherlands. The success of Silicon Valley is often attributed to a self-reinforcing 'cluster effect'.[17] Groups of like-minded companies attract talent, venture capital and new companies, with the appeal being amplified by the exchange of people, knowledge and ideas within the cluster.[18] Science is subject to the cluster effect as well: a high concentration of talent helps to attract new talent.

Within the Netherlands some regions specifically promote themselves as clusters of high-tech, life sciences, energy, sustainability or creative industry businesses.[19] On a global scale however the Netherlands is one single compact and exciting knowledge cluster, situated close to other major clusters in Germany, France and the United Kingdom.

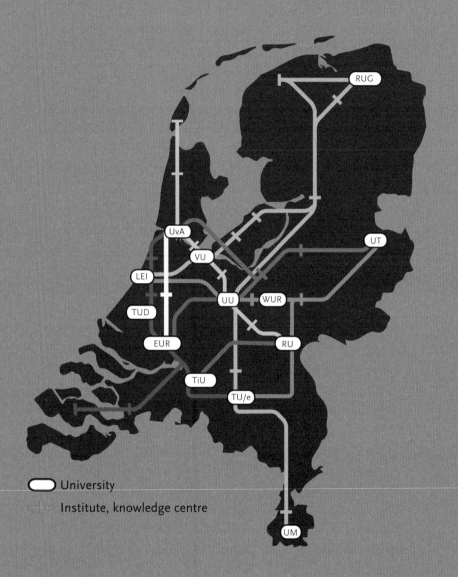

University
Institute, knowledge centre

A COUNTRY OF CLOSE CONNECTIONS

Being a small country, the Netherlands has created a close-knit web of col-
laborating and integrated institutions—a closely connected network of thirteen
universities and dozens of research and knowledge institutes. No researcher
in the Netherlands lives more than two hours away from the centre of the
country.

The Netherlands is not only a compact country, it has also invested heavily in creating excellent connections.

That starts with physical connections and a highly developed network of roadways, railways and flight routes, which make it easy for researchers to visit colleagues elsewhere in the country and abroad.

Measured by kilometres of track per square kilometre, the Netherlands has the fourth densest railway network in Europe.[20] The country has the highest density of roads in Europe after Belgium.[21] The World Economic Forum has described the Netherlands' infrastructure as the third best in the world, after the city states of Hong Kong and Singapore.[22]

A researcher at the University of California in Berkeley who wants to visit a colleague in Santa Cruz, 130 kilometres away, has to choose between sitting hours in traffic or travelling for almost four hours by train. A physicist in Amsterdam who would like to consult a colleague in Eindhoven can travel back and forth in the same time. No researcher in the Netherlands has to travel for more than two hours to reach the centre of the country, which partly explains the extensive collaboration and consultation that takes place at the national level.

There are also good international connections. Only Frankfurt airport has more direct worldwide connections by air than Schiphol.[23] No country in the European Union has a better digital network than the Netherlands.[24] Amsterdam is one of the largest Internet hubs in the world and, thanks to the network of SURF, a public organisation dedicated to enhancing IT facilities in research and education, Dutch researchers can share ever larger volumes of data at ultra-high speeds.[25]

Last but not least, there are exceptionally good connections in an organisational sense. The science system in the Netherlands has invested in making connections at every level in the form of networks and personal contacts through which knowledge and facilities are shared. Compared to other countries, the Dutch system of scientific research and education is extremely tight-knit.

By virtue of all of these connections the Dutch universities, institutions and companies strengthen and complement one another, which ensures that the science system as a whole is greater than the sum of its parts.[26]

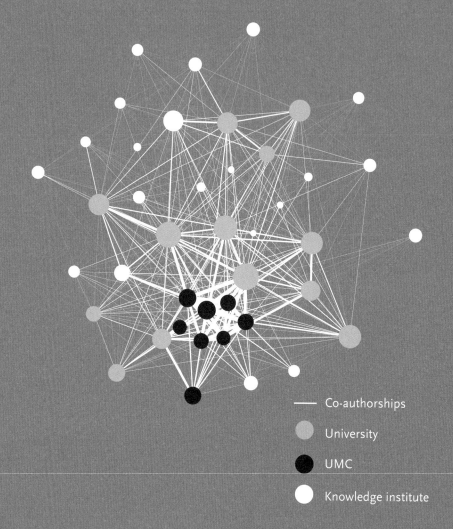

— Co-authorships

● University

● UMC

○ Knowledge institute

A WEB OF INTENSIVE COOPERATION

There is intensive collaboration between individual researchers and research institutes in the Netherlands. Together they form a single close-knit network. The figure shows the co-authorship of academic articles by researchers from universities, university medical centres and research institutes between 2008 and 2017. The thicker the line, the larger the number of joint publications.

Source data: SCImago Journal & Country Rank (www.scimagojr.com) on the basis of Elsevier's Scopus database.

Science in the Netherlands also benefits from what is often referred to as 'the polder model': consultation, consensus, cooperation and bottom-up innovation are all part of a unique scientific culture.

Although the circumstances have constantly changed in the course of history, traces of this Dutch 'polder model' can be found going back a long way. Over the centuries the Netherlands seems to have developed its own approach to meeting challenges and organising effective bottom-up responses to them—the construction of reliable systems to protect low-lying polders against flooding is just one example.[27]

The Academy's former president Robbert Dijkgraaf once formulated the Dutch approach as follows: *"A crucial success factor [of Dutch science] is the strong interconnectedness, the accursed consultative structure and polder mentality, which is endured every day in countless grey meeting rooms with researchers drinking bad coffee from plastic cups. It [..] gives [..] a unique coherence to our country."* [28]

One way in which the polder model is reflected is in the fact that Dutch researchers are accustomed to excelling in 'friendly competition'.[29] Scientists compete, but also collaborate, with each other at every level. It is a tradition that determines the interaction between individual researchers, but also between research groups, institutes, disciplines and other research organisations.

Friendly competition means that researchers assist one another where necessary and seek cooperation with colleagues from other universities. Disciplines or sub-disciplines join forces in national research schools to train young researchers and initiate nationwide research programmes. Top researchers from different universities work together intensively, for example in the special Gravitation research funding scheme.[30]

In the world of science, the polder culture and organisational capacity is reflected in multidisciplinary research schools and, on a larger scale, in collaborative programmes of research funders.[31] There is also cooperation with and between non-university knowledge institutes and civil society partners. To mention just one of many examples, scientists in the humanities collaborate intensively with museums and public research institutions, including the Netherlands Institute for Art History and the National Association for Cultural Heritage, in research performed by the Postgraduate School for Art History.

More than ever the Dutch polder model also encompasses parties from the private sector. Private laboratories (such as the renowned Philips NatLab) have generally abandoned fundamental research or play only a minor role in it.[32] Innovation increasingly happens in the context of public-private partnerships between companies, academia, the government and civil-society organisations. These alliances encompass many types of companies, from small start-ups and family-owned enterprises to multinational corporations.[33]

Another good example of the unique Dutch polder model is the existence of private charitable funds that raise money for specific biomedical research, such as for the treatment of diseases including cancer, cardiovascular diseases, and Alzheimer's. These charitable funds are a useful supplement to the funding of scientific research: the Netherlands Cancer Institute owes its scientific impact in part to the fund-raising efforts of the Dutch Cancer Society. Equally important, however, is the fact that the charitable funds have brought a unique involvement of patients and their families in science. Nowhere is the connection between patients and researchers as close as it is in the Netherlands.[34]

The compilation of the National Research Agenda has been another perfect illustration of how researchers, the general public, professionals and representative organisations can be brought together. Starting in 2015, in a unique experiment, 12,000 questions submitted by both citizens and scholars were distilled into two dozen research themes; in coming years, researchers from very diverse disciplines and a wide range of organisations and businesses will look for answers to important scientific, societal and economic challenges. It is yet another example of science shaped according to the Dutch polder model: bottom-up, open and driven by collaboration and friendly competition.

A question of trust

Mutual trust is crucial for every one of these forms of collaboration. According to the American political scientist Francis Fukuyama, trust is key to producing social capital that allows societies to develop.[35]

A society with trusted civil-society organisations derives self-organising capacity from that trust. Its businesses and organisations can grow, become more professional, engage with each other, make agreements and collaborate.

In Fukuyama's terms, the Netherlands is a 'high-trust society' and the country's researchers together form a 'high-trust research community'. One aspect of that trust is that Dutch universities are given a relatively large degree of freedom by the government to appoint professors and to plot their academic course.[36] In countries like Germany and France, government approval is needed for many appointments.

In recent years the Dutch government has again tied part of its budget for universities to specific policies. Universities have had to sign 'performance agreements' that require them to bring more focus to their research and to gear it more to the needs of business and society. This development suggests a weakening of trust.

Mutual trust is critical for fruitful and sustained scientific cooperation, whether it is between public research institutes themselves or, more particularly, between public institutions and business. Only in relationships that are built on trust can companies share their expertise and can public knowledge be quickly translated into useful applications to society's benefit.

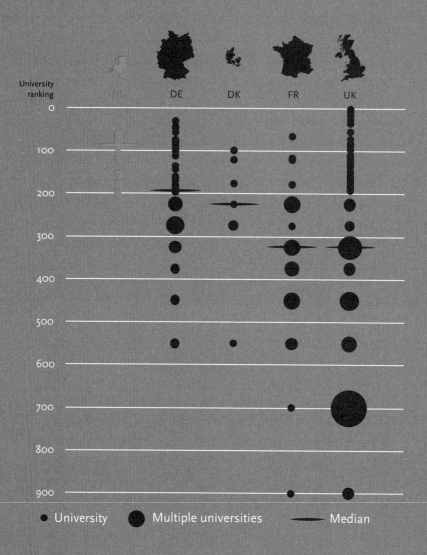

University ranking

| | NL | DE | DK | FR | UK |

- 0
- 100
- 200
- 300
- 400
- 500
- 600
- 700
- 800
- 900

• University   ● Multiple universities   — Median

HIGH QUALITY ACROSS THE BOARD

None of the Dutch universities fall outside the top 200 in the Times Higher Education ranking of the world's 980 best universities: Dutch universities are all of a high standard, in contrast to those in, for example, the United Kingdom, which are well represented at both ends of the table. The horizontal lines represent country medians—the point at which as many universities in that country rank higher as lower.

Source: Times Higher Education World University Rankings 2016-2017.

Some countries seek strength in strict selection and the concentration of funding and talent. Dutch science also has its selection processes, but it has always known an egalitarian tradition as well: there is an aversion to hierarchies and excessive differences.

Former Academy president Hans Clevers regarded the typical Dutch lack of respect for hierarchy as a pillar of the success of research in his country because it gives talented young scientists the chance to climb to the top. "When I [as a professor] suggest a plan, there is a discussion in my research group: is it really a good idea? In other countries, young researchers are often in awe of the professor; they will not contradict him, do not wish to disappoint him," he once said in an interview.[37] A fundamental characteristic of science is the ability to challenge established ideas. *"Science is the belief in the ignorance of experts,"* as Richard Feynman once put it.[38]

In Dutch universities and institutes a culture of friendly competition prevails, with ingredients such as cohesion, mutual assistance, discussion about research and cooperation inside and outside the group, but also an openness to criticism, to newcomers and to new directions in research.

The Dutch research system is also egalitarian in other respects. The possibility of studying at a good university is not hampered by strict selection based on grades or high tuition fees. In principle, every student can be inspired by the lectures given by Dutch top researchers. There is not as much discrepancy between the budgets of various universities as there is in countries like the U.K. or the U.S.

As a result, the standard of Dutch universities is high across the board. The figure on the opposite page, based on the Times Higher Education worldwide ranking of universities, shows how evenly quality is divided in the Netherlands.[39]

All 13 Dutch universities are now in the top 200 in this ranking.[40] No Dutch university comes close to matching Oxford, but as a country, the Netherlands ranks highest worldwide. The median position of Dutch universities in the 2016 ranking was 86th; that of universities in the U.K. was below 300th.

Certainly, selection based on excellence can attract top-class students, and large individual grants such as those provided under the 'Vernieuwingsimpuls' (Innovational Research Incentives Scheme) can strengthen the Dutch science system.[41] But one should never forget that the Dutch tradition of equality and a broad research base are integral parts of the country's strength as well.

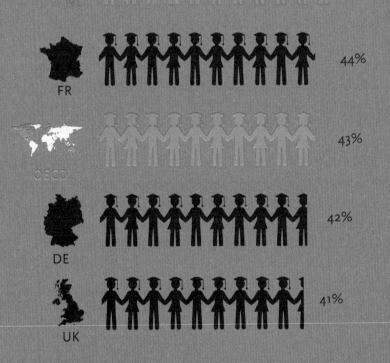

DK — 52%
NL — 48%
FR — 44%
OECD — 43%
DE — 42%
UK — 41%

INTERNATIONAL CO-AUTHORSHIP

Dutch researchers collaborate a lot with colleagues in other countries, even compared with researchers in other countries. The Organisation for Economic Cooperation and Development (OECD) investigated the percentage of articles in peer-reviewed scientific journals in which the author list includes at least one author working in another country.

Source data: OECD en SCImago Research Group (CSIC) on the basis of Elsevier's Scopus database.

One of the reasons why Dutch research has such a high citation impact worldwide is the country's international orientation. Dutch researchers are not only proficient in other languages, they also collaborate relatively frequently with colleagues in other countries.

Almost half of all Dutch research publications have at least one co-author working in another country, often Germany, Belgium or one of the Scandinavian countries.[42] One in twelve European research projects financed from Horizon 2020, the European Commission's Eighth Framework Programme for Research, is coordinated by a Dutch researcher.

Many Dutch researchers spend at least part of their career working abroad. Conversely, in 2013 the proportion of Master's students from other European countries studying in the Netherlands was two-and-a-half times greater than the European average.[43]

The international orientation of the Dutch academic community is promoted by the traditional practice of NWO, the Dutch granting organization, to have research proposals evaluated by international referees whenever possible. Visitation committees that assess university research programmes are also highly international in composition. Accordingly, it is commonplace for Dutch researchers to measure their work against that of colleagues in other countries.

The growing practice at Dutch universities to teach some courses (as well as conducting research) in English also strengthens the international orientation of science in the Netherlands.[44]

These success factors all contribute to the fact that research conducted in a small country like ours can measure up to that of larger countries. But, as is so often the case, past results offer no guarantee for the future. We will have more to say about that later.

3 Meanwhile at the neighbour's

To put the state of Dutch science into perspective, it is useful to look at what is happening in countries close by. A glance at science systems in Germany, France, the United Kingdom and Denmark, some of the Netherlands' neighbours, can serve to highlight some interesting differences.

Germany: solid
investor in innovation

Germany has a very strong science system, which benefits from the certainties of long-term, relatively heavy investment by the government and the private sector in research and development.

In addition to the universities, the German science system includes a number of research organisations such as the Max Planck Gesellschaft, the Helmholtz Gemeinschaft and Leibniz Gemeinschaft for research facilities and institutes, and the Fraunhofer Gesellschaft for applied research.

Compared with the Netherlands, institutes like Max Planck are more generously financed by the government than the universities, a distinction that has been partially repaired with the *Exzellenzinitiative*.[45]

The links between the institutes and universities in Germany are not as close as they are in the Netherlands, partly because of the greater distances. Tighter control and a stricter hierarchy mean that German universities are less flexible than those in the Netherlands. The differences between universities in terms of quality are greater.

In contrast to the Netherlands, during the recent economic crisis, public and private investment in research and development in Germany did grow to 3% of gross domestic product (GDP), the target set by the European Union in Lisbon in 2000.[46] Germany's aim is to increase the level of investment to 3.5% in 2025.[47]

The generous German funding is reflected in, among other things, large grants designed to attract outstanding talent from abroad. Every year the German government funds attractive positions at German universities, worth five million euro each, via the Alexander von Humboldt Stiftung. Next year, two of the five Alexander von Humboldt Professorships will go to leading Dutch researchers, offering them better opportunities in Germany.[48]

Private foundations, such as the Volkswagen Stiftung and the Robert Bosch Stiftung, play an important role in funding research in Germany.[49] Through these foundations companies can make a major contribution to promoting cutting-edge research and to forging links between science and industry.

United Kingdom,
champion of
competition

The universities of Oxford and Cambridge often come out on top in European rankings of universities. And they are not the only excellent universities in the United Kingdom.

This partly explains why 21% of the prestigious grants awarded by the European Research Council go to that country and why its universities attract talent from around the world. But like the United States, the United Kingdom also has numerous universities that perform far less well. There is a wide disparity between the peaks and the valleys.

The Research Excellence Framework (REF), a national cycle of quality assessment that is far more intensive than that faced by universities in the Netherlands, plays a major role in the British system. The impact of academic research at every university is evaluated every seven years and the ensuing table plays a role in determining the funding of the universities.

The REF has a lot of influence on the activities of universities. The intense competition between universities promotes excellence and diversity, but it is not an incentive for cooperation. The REF can even disrupt research: every seven years, just before the evaluation, a sort of transfer market for good researchers opens up.[50] There is also a clear pecking order among researchers within universities. The emphasis is more on competition and hierarchy than on collaboration.

The U.K. has seven Research Councils that play an important role in publicly financed research, for example in terms of initiating research programmes. The Research Councils were a source of inspiration for the NWO's four domains, although it should be noted that NWO also encourages cooperation across the boundaries between its domains.

Finally, the private and independent Welcome Trust, the largest philanthropic institution supporting scientific research in the world after the Bill & Melinda Gates Foundation, is an important player in the field of research.

The United Kingdom's pending exit from the European Union could have harmful consequences for research in Britain because European research subsidies may disappear and there will be greater obstacles to recruiting talent in the EU. The British government has announced that it will increase the national science budget by billions of euro from 2020 in order to counter the negative effects of Brexit.[51]

France: strong central
government control

ATTENTION à la MARCHE

In France as well there are major disparities in the standards of universities. The prestigious *Grandes Écoles*, which are concentrated in Paris, attract talent from the entire country, which makes it difficult for regional universities to climb in the ranking. Generally speaking, French science as a whole has a moderate rating in international terms.

The system in France is a lot more hierarchical than in the Netherlands and there is a lot more control. Many appointments by universities require the approval of the national government.

An important role is played by the *Centre National de la Recherche Scientifique* (CNRS). University researchers who have secured a position at the CNRS devote little time to teaching anymore. The universities appoint young researchers to tenured positions with extensive teaching responsibilities often soon after they have obtained a PhD.[52] This does mean that young researchers are less reliant on securing a succession of temporary post-doc contracts than their Dutch counterparts.

45

Researchers have few opportunities to secure external financing and consequently have little capacity in terms of research materials and temporary research staff. On the other hand, they have greater freedom to choose their own subjects for research over a longer period. With tenured appointments and the minor role of external financing, French researchers have a lighter workload than researchers in the Netherlands.[53]

French companies such as Servier and Sanofi (pharmaceuticals), AXA (insurance), Dassault (aircraft) and Louis Vuitton (fashion) invest in institutes and fund professorships. However, there are few organised public-private partnerships along the lines of those in the Netherlands.

Like the United Kingdom, France devotes a lot of attention to science communication and education.

Denmark: growing investment in a clear strategy

In our comparisons of neighbouring countries, Denmark and the Netherlands often scored similarly. Both are small countries with remarkable records in terms of research. One notable difference is that, like Germany, Denmark has continued to invest ambitiously in science and innovation over the last few years.

As a result, Denmark is one of the few EU member states that meets the Lisbon target of investing 3% of GDP in research. Danish companies and the Danish government invest significantly more than their Dutch counterparts.

Denmark has a clear vision of the importance of science, technology and innovation. The country has translated that vision into equally clear ambitions, which are pursued and financed under the guidance of a single, compact ministry. The ministry has set out transparent programmatic lines, clearly delineating strategic lines formulated at the national level from research that is driven from the bottom up. In contrast to the trend in the Netherlands, Denmark consciously continues to provide generous support for smaller research projects in addition to strategic programmatic research. The Danes also invest more in technology and innovation.

Denmark attracts more top senior researchers from other countries than the Netherlands.

Around 2008 the number of Danish universities was reduced and national institutes were integrated into those universities. In 2012, the Danes formulated a new national innovation strategy entitled 'Denmark—a nation of solutions'.[54]

As in Germany, private foundations such as the Carlsberg Foundation, the LEGO Foundation, the Lundbeck Foundation and the Novo Nordisk Foundation help to support fundamental research and to attract talented researchers.

4 The base is eroding

Dutch scientific research owes its
success to some unique characteristics.
But that does not mean such success is
guaranteed.

While reaching for higher tops, the country
sometimes neglects its foundations.
Higher peaks are resting on a crumbling
base.

Public and private investment is lagging

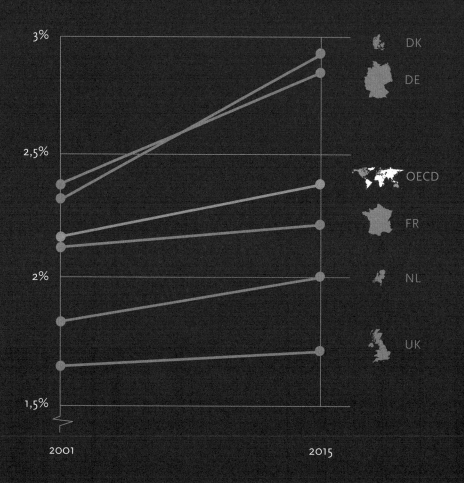

THE NETHERLANDS INVESTS LITTLE IN RESEARCH

For years the Netherlands has invested less than other countries in research and development. Public and private investment together came to just over 2% of GDP in the Netherlands in 2015, whereas Sweden, Austria, Denmark, Switzerland, Germany and Finland all invested around 3%.

Source data: OECD Main Technology and Science Indicators: GERD as a percentage of GDP.

It is a recurring refrain in every analysis of science in the Netherlands: the country invests too little in research and development.

In 2000 the EU leaders decided in Lisbon that three cents of every euro we earn should go to research and innovation, knowing that such an investment is crucial for future economic development. The Netherlands is making scarcely any progress towards meeting that goal, even though other countries have made clear advancements.

Investment by business and the government has increased in recent years, but only barely.[55] Public investment in the Netherlands is slightly higher than the EU average, while private investment, which accounts for roughly half of the total, is slightly lower than the EU average. All in all, the Netherlands struggles to reach 2% of GDP.

Meanwhile, other European countries are already doing a lot more. Sweden, Austria, Denmark, Switzerland, Germany and Finland are all at around 3%—to say nothing of Japan and South Korea. Every year, the Netherlands is slipping further behind.

Part of the problem is that when it comes to research, 'investment' is too often regarded as 'expenditure'. The models used by the Netherlands Bureau for Economic Policy Analysis (CPB) do not translate investments in knowledge and innovation into future economic growth, even though we know that a large proportion of the economic growth generated in the past was the product of research investments.[56]

Universities can spend less on research

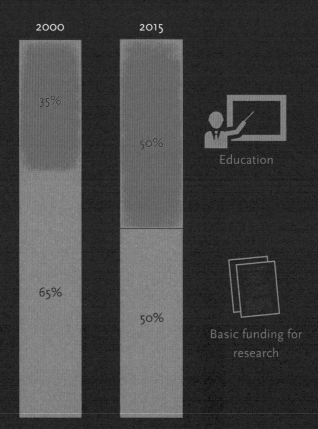

2000   2015

35%

50%   Education

65%

50%   Basic funding for
             research

SHRINKING SHARE OF DIRECT FUNDING GOES TO RESEARCH

The amount that universities receive directly from the government has barely risen
since 2000, while the rise in student numbers means that more money is going
to teaching and less to research. In the formula used by the Ministry of Education,
Culture and Science to allocate funds to universities, the share going to research fell
from 65% to 50% between 2000 and 2015.

Source: Letter from the Minister of Education, Culture and Science to the House of Representatives,
24 May 2017.

The absence of new investment in research, technology and innovation is all the more pressing because institutions are having to finance additional activities from a budget that has remained more or less stable.

A budget that has remained steady now has to be used to teach a growing number of students and PhD candidates; to create research programmes for economic top sectors; and to set up multidisciplinary research programmes for addressing urgent societal challenges.

The close relationship between research and teaching, one of the Netherlands' strengths, has also masked an insidious shift in spending within Dutch universities. That shift has gradually drained the budgets for fundamental research, the foundation on which the entire science system is resting.

Dutch universities are funded for a large part on the basis of student numbers. This direct government funding (the 'first flow of funds') finances both teaching in the universities and the broad base of university research.

Between 2000 and 2015, the number of students rose by 54%, while direct government funding only grew by 12% during that same period.[57] Whereas 65% of the university funding used to be intended for research, that share has now shrunk to just 50%.[58]

Over the same time frame, the number of PhD candidates rose even faster than the general student population—PhD student numbers doubled between 2000 and 2015.[59] But PhDs are not budget-neutral for universities: on balance, supervising and providing the infrastructure for PhD research costs more than the income generated by subsidies for projects or the so-called 'PhD bonus'. In other words, the teaching of PhD students accounted for a steadily growing share of the direct government funding.

We conclude that money that should go to the broad base of fundamental university research is in practice being used to pay for growing teaching activities. That shift is undermining research, and it threatens the traditional interwovenness of education and research in the Netherlands.

External funds lay claim to the base

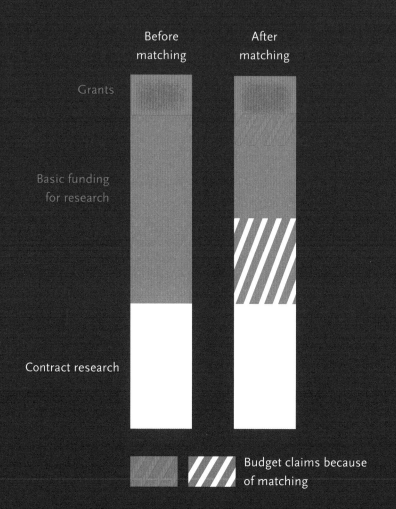

Before matching · After matching

Grants

Basic funding for research

Contract research

Budget claims because of matching

EXTERNAL FUNDING LAYS CLAIM TO THE BASE

Universities have come to depend greatly on grants (the 'second flow of funds') and contract research (the 'third flow of funds') to pay for research. Many external funders however demand that universities match the grants they receive with funds of their own (the 'first flow of funds'). Accordingly, the second and third flows of funds now lay substantial claims on the budget that used to be intended for basic research.

Source: Spinning plates; funding streams and prioritisation in Dutch university research. Rathenau Institute, 2016

It is not only the government that contributes to the increasing strain on the research base. The growing pressure also stems from external funders and from new demands being made by society.

External funders increasingly insist that universities match the grants they receive with substantial sums from their own financial research resources. The matching funds enable the funders to multiply the impact of their contribution and ensure concordance with other research. They make efficient use of infrastructure that has already been financed from the basic funding. This happens with NWO programmes, with grants from businesses and charities and with collaborative projects funded by the European Commission.

Accordingly, matching lays growing claims on government funding that was originally intended for the university's own research priorities and its basic research infrastructure. For every euro in external funding they receive, universities have to provide 75 cents on average from their own funds.[60] External funding can therefore become a mixed blessing—for a small faculty, a large grant can even turn into a minor disaster because matching requirements force it to suddenly scale down other research or divert funds originally allocated to recruiting top researchers. Many universities can no longer afford to recruit their own PhD candidates.

The conflicting interests clash in the offices of deans, who often feel they are forced to keep numerous plates spinning.[61] [62]

All of this pressure is causing more and more universities to insist that professors and researchers spend parts of their time recouping their entire salaries by raising (increasingly scarce) external funds. Even young researchers, who have five years to prove themselves and qualify for tenure, are assessed on their fund-raising capabilities.

As a result, researchers experience more competition and higher workloads, and their room for collaboration and friendly competition is shrinking.

These factors undermine the basis of the Dutch research system from within.

The steadily growing pressure on the funding of 'free and unfettered' research[63] in the Netherlands is the outcome of numerous factors.

Another one of those factors is the disappearance of the Economic Structure Enhancing Fund (FES), from which hundreds of millions of euro were invested in research between 1995 and 2010. The final FES programmes came to an end in 2015. Their termination removed the financial base for thousands of PhD positions at universities.

The introduction of a new industrial policy in 2011 has also had a negative impact on the budget available for fundamental research.[64] Since then NWO has to earmark a sizeable part of its budget for grants for economically important 'top sectors' and thematic research programmes. Much of that money was taken from the budget for non-thematic, researcher-driven studies. Growing numbers of researchers are now competing for shrinking budgets for that type of research, which is inefficient and demoralising.[65]

All in all it is a recurring pattern: a government finding the funds for new research policies by taking them out of other parts of the research system, typically from its foundation. The word often used for this practice is 'reallocation', but a less reverent term might be 'cannibalisation', because the process is taking ever more resources from the broad base on which the Dutch research system is built.

More tasks, less funding

While researchers have to work harder to survive, the demands that society is placing on them are increasing.

More than before, politicians and the public want universities to devote attention to how knowledge can be applied and monetised—and rightly so.

Understandably, they also insist that researchers should engage in disciplines other than their own, form multidisciplinary teams that can address societal challenges, and invest time and effort in educating and inspiring other members of society.

The government wants university education to radically renew itself. And naturally it wants every scientist to respect ethical and other guidelines governing scientific research, even if those rules make conducting research more difficult, more cumbersome or more expensive.[66]

For researchers, however, all of this raises the question of how they can meet all of the demands while simultaneously having to perform ground-breaking research and raise external funding.

Too much pressure on the system will go at the expense of motivation and quality; it could lead to risk aversion, and hence to less innovation. Excessive fluctuations in funding can endanger the continuity of research, not only at universities but also at research institutes.[67]

Too much pressure could also transform quality indicators into perverse incentives and prompt disproportionately strategic behaviour in order to meet those incentives. In recent years, the emphasis on scientific integrity has rightly increased.

The Dutch science system being extremely efficient and generating lots of citations per euro invested may seem to be all good news. It could signal that Dutch research money is being used very effectively. Yet it could also hide an implicit risk: The Netherlands may have designed the system in such a way that its researchers are focussing too heavily on very specific indicators.

Bibliometric indicators of quality are regarded by some as perverse incentives—incentives that are so powerful that they wrongly push other important objectives of research into the background. Excessive pressure to score on narrowly defined performance indicators can, for example, lead to scientists cutting corners or concentrating too much on securing media publicity.

We do not believe that publications, citations and grants in themselves constitute perverse incentives that are gnawing at the roots of science, but there is clearly an inherent risk to concentrating too much on bibliometric indicators.

Some universities are already trying to restore the balance by giving more attention to the social significance of research and researchers' willingness to collaborate in the recruitment and appraisal of staff.[68] The Standard Evaluation Protocol (SEP) for scientific research, drafted by the Academy, NWO (the Dutch granting organization) and the VSNU (the association of Dutch universities), has also been revised in that direction. These are trends that deserve support.

Another type of perverse incentive worth mentioning here is a self-reinforcing 'Prof. Scrooge McDuck effect',[69] in which a small number of researchers accumulate large amounts of research funds while many good researchers around them can scarcely survive.

This can lead to a sort of pyramid structure in which a few researchers at the top, with few teaching responsibilities, are surrounded by large groups of teachers who are no longer able to do much in the way of research. It also leads to some excellent research groups being disproportionately squeezed.

This is a downside of success that flies in the face of the Netherlands' egalitarian tradition and could eventually undermine the broad base of the Dutch science system.

Great talent is being
lured away

 DEPARTURES

TIME	DESTINATION	FLIG
12:39	LONDON	CL
12:57	SYDNEY	UQ5
13:08	TORONTO	IC5
13:21	TOKYO	AM
13:37	HONG KONG	IC5
13:48	MADRID	EK3
14:19	BERLIN	AM5
14:35	NEW YORK	ON
14:54	PARIS	MG5
15:10	ROME	RI5

The shrinking budget for free university research and more strident competition for shrinking funds threatens to weaken the Netherlands' appeal to talented scientists.

Some of that talent is to be found abroad and it is already proving difficult to attract it to the Netherlands. The Breimer Committee concluded that the Netherlands is no longer able to compete in attracting top senior researchers in chemistry and physics.[70] The rector of Delft University of Technology has said that his university finds it hard to attract talented young scientists because it is unable to match the start-up packages that researchers are being offered in other countries.[71]

The most important reservoir of talent is therefore the Netherlands itself, but Dutch researchers are also increasingly exposed to temptation from elsewhere. Countries such as Germany, the United Kingdom, United States and Switzerland have greater financial resources and offer good positions to talented Dutch scientists. A senior researcher in Switzerland can earn up to twice as much as his counterpart in the Netherlands.[72]

Several times a year the newspapers report on yet another exceptionally talented Dutch researcher moving abroad. Each case is unique and it is difficult to draw general and firm conclusions, but scientists in the Netherlands know that many of their highly talented peers are being lured to other countries. Spinoza Prize-winner Bert Weckhuysen issued a public warning that his young researchers tell him they no longer want to remain in what they perceive as a 'rat race'.[73] This general trend will be analysed in an advisory report to be published by the Academy later this autumn.[74]

5 Nurture Dutch strengths

A strong and unique science system that is gradually eroding its broad base in order to reach new heights—that is the pattern that keeps emerging. The question is: where do we go from here?

This final chapter presents some ideas for restoring the foundations on which the Dutch science system was built.
The goal remains the same: use Dutch openness, collaboration, trust and organisational capacity to keep reaching the top in research, and apply more of that knowledge than ever before.

THE NETHERLANDS AS A HIGH PLATEAU

The research landscape in the Netherlands can be described as a high plateau
in relation to the countries around it. The height of the countries in this map
reflects the median position of their universities in the global Times Higher
Education ranking. The Netherlands is on top thanks to past investments, but
will have to gradually step up its research funding to avoid sinking back.

Source: Times Higher Education World University Rankings 2016-2017.

Like others have done before us, we call for more investment in Dutch research and innovation in both the public and private domains. The Netherlands should follow the example of countries around us and quickly take steps to meet the target of investing at least 3% of GDP in research every year.

Strengthening the base does not call for a large incidental financial injection in new policy, possibly followed by spending cuts a few years later. On the contrary: additional shocks to the system must be avoided. There should preferably be a lengthy period of modest but steady growth, for example by establishing a special fund for government investment in infrastructure, education and research.[75]

The steady growth should be used to restore and strengthen the base across the board: universities and research institutes (the first flow of funds), competitive financing (the second flow of funds) and programmes dedicated to public and public-private collaborative partnerships. The investment agenda[76] produced by the Knowledge Coalition,[77] with its specific proposals for investment in talent and infrastructure, would form a good starting point.

For the longer term the government, the universities and NWO should together reflect on changes to performance agreements, allocation models and financial instruments in the first and second flows of funds, since they are the cause of the pressure felt by university managers, faculty deans and researchers.

The universities and research institutes themselves should be involved in that discussion as well. They can ensure that additional investments will strengthen the base rather than increase the pressure even further. It might be necessary to design new models for the internal allocation of budgets, with new incentives and new human resources policies. Strengthening the base also implies: more room for collaboration, for research on socially relevant themes, for communication and interaction with society, for knowledge application, for safeguarding scientific integrity and for preventing sloppy science.[78]

It is not only the government that should invest more in research and development; companies, large and small, should be encouraged to invest in innovation and in establishing a broader base for Dutch scientific research.

Dutch companies lag behind enterprises in other countries in international comparisons of spending on research and development: their expenditure is 13% below the EU average and 48% below the average of the OECD countries.[79] The existing methods of stimulating investment by the business community therefore need to be strengthened. One instrument we would like to draw special attention to is that of private foundations and trusts.

We have seen how the science systems in Germany, the United Kingdom and Denmark benefit from contributions from large private funds which add support for research that has no direct economic application. They give universities and institutes more room to hire talent, support creative new ideas, and contribute to a healthy base of fundamental scientific research.

The Netherlands already has charities that raise funds for medical research. Some universities have established funds that support specific projects. But compared to other countries, the Netherlands has hardly any large funds dedicated to science as such.[80]

We call for new incentives to alter that situation. As in Germany and Denmark, the Dutch government could provide greater incentives for businesses to establish funds to support research in the Netherlands, including basic research.[81]

These funds would not only be able to provide more money, but could also contibute by connecting science with private partners and by increasing support for science in society as a whole.

Strengthen
connections and
cooperation

The National Research Agenda ('Nationale Wetenschapagenda') has been a perfect example of polder science: bottom-up, open, and focusing on collaboration in friendly competition.

We consider the Research Agenda's overarching themes ('routes') to be good starting points for organising collaboration among researchers from many disciplines to address major scientific, societal and economic themes.[82] It is vital that the Agenda now can continue along the path that has been taken: the research themes must be allowed to come to fruition.

Cooperation should also be promoted between disciplines that have not traditionally worked closely together in the Netherlands, partly as a result of a unique historical distinction between 'general universities' and 'universities of technology'. Outside the Netherlands, the term 'general university' requires explanation.

Because of this distinction, Dutch engineers interact less with scientists in biomedicine, the social sciences and the humanities than in other countries, and most general universities do not have engineering departments.[83] That is one of the explanations for the relatively wide gap between fundamental and applied research in the Netherlands.[84] Investments in stronger connections between general universities and universities of technology would help to narrow that gap, while affirming the distinction would be a step in the wrong direction.

In line with the National Research Agenda and its integration with the top sectors, collaboration with and between universities, KNAW and NWO institutes, TO2 institutes and public knowledge institutes should be promoted.[85] The same applies for the many forms of public-private partnership between universities, research institutes and innovative companies. Further regional and national clustering will make the Netherlands more effective and more attractive in both scientific and economic terms.[86] The growing number of start-ups emerging from universities should be nurtured.[87]

The interweaving of research and teaching, a key Dutch strength, should be protected; top researchers are great sources of inspiration for students, and outstanding teachers must be able to keep conducting their own research.

A 'Harvard on the Rhine'—a single university that can stand out from the rest thanks to extra resources—is not part of our vision. Such an approach would go against the egalitarian roots of the Dutch science system, and would more likely weaken that system rather than strengthen it.[88]

Utilize
self-organising
capacity

At NWO, the opportunities for fundamental research have sharply declined in recent years because of shifts in funding.[89]

A reinforced second flow of funds should include greater possibilities for collaboration. It should provide financial space for multidisciplinary programmes in which researchers from the natural sciences, the social sciences and the humanities can work together with peers from the technical sciences and with companies, and in which they can jointly address complex scientific and societal challenges.[90]

Through open calls for programme proposals with budgets of between 1 and 2.5 million euro, researchers should once again be given explicit opportunities to initiate and design their own multidisciplinary, thematic research programmes.

Such programmes would nurture innovation and collaboration within and across disciplines, and thus also promote research on the themes in the National Research Agenda.

75

Give talent breathing space

When strengthening the science base in the Netherlands, it will be crucial to create breathing space for talent. Talent keeps the system afloat.[91]

Promoting talent has to start with instilling curiosity and an interest in scientific research in school children, including encouraging them to choose an education in science or engineering. Promoting talent also means removing invisible obstacles that keep some women and young researchers with migration backgrounds from reaching every level of scientific research. The Netherlands still lags behind internationally in terms of the proportion of female scientists. A national programme to promote diversity is desirable.[92]

Promoting talent also implies increasing the country's appeal to top scientists. Leading Dutch researchers who are drawn away to attractive positions elsewhere regularly express their disappointment in an underfunded Dutch research climate. [93]

At the same time, universities should display more creativity in attracting top talent, students and young researchers from abroad. An interesting idea in that context is to accentuate the coherence of Dutch universities in international marketing by strengthening the concept of the 'University of the Netherlands'. Bibliometric analyses show that the Dutch universities are more integrated than the ten public, reasonably autonomous campuses that together make up the University of California.[94]

To be sure: we are not calling for the thirteen Dutch universities to merge. What we do advocate is joint international promotion and cooperation in attracting talent from abroad: students, PhDs, post-docs and senior researchers. By working together Dutch universities might, for example, find it easier to offer jobs to the partners of researchers—in the United States one in three researchers has a partner who also works at a university.

Having been recruited, the talented scientists will have to be shown the ropes of the unique procedures in the Dutch science system, which to outsiders can easily be a source of amazement.

Strengthening the Dutch science base also calls for the rebuilding of trust because a number of traditional relationships have been tested in recent years.

This applies, for example, to the researchers and the research institutes who have become disillusioned with politicians due to inadequate budgets, expanding tasks and increased steering from above; who feel they have to meet growing demands while facing more uncertain career prospects; and who are asked to collaborate with one another while at the same time being forced into more fierce competition.

Rebuilding trust also applies to society—partly because of the growing 'rat race' in science, the general public is hearing more stories about sloppy or unethical research, and more often questions the roles and interests of researchers.

In line with the government's 'Vision for Science', we advocate research subsidies that span longer periods, with thorough *post facto* accountability replacing frequent and detailed interim controls. This would help scientists to devote their time to what they are good at: doing research and teaching.

We also like to mention some inspiring initiatives that have helped build trust between science and business: the MIT Energy Initiative in Boston[95] and, in the Netherlands, the Industrial Partnership Programme[96] of the former FOM. Both have shown that it is possible to forge enduring partnerships between science and business without making concessions to the quality and the impact of research.

We feel it is very important to build trust in these and other ways.[97] For that, in the spirit of the Dutch polder model, consultation and collaboration by all of the parties concerned will be essential.

On the basis of this vision, we and the other members of the Academy's board will endeavour to prolong the success of a very Dutch system of science, and in so doing contribute to the country's success for years to come.

Notes

1 Gross Domestic Product 2016 (based on purchasing power parity). Source: IMF, World Economic Outlook Database, April 2017.

2 Post Statistics 2013, Press release from the Port of Rotterdam Authority.

3 CBS, March 2017.

4 Articles in every area of science, published in 2014 in the more than 21,000 peer-reviewed journals with English-language abstracts in Elsevier's Scopus database. Source: SCImago. (2007). SJR—SCImago Journal & Country Rank. Retrieved July 21, 2015, from http://www.scimagojr.com.

5 Citations of articles in every area of science, published between 1996 and 2016 in the more than 21,000 peer-reviewed journals with English-language abstracts in Elsevier's Scopus database. Source: SCImago. (2007). SJR—SCImago Journal & Country Rank. Retrieved July 21, 2015, from http://www.scimagojr.com.

6 These figures could have been distorted by the selection of journals with at least summaries in English. That is not the entire story however: measured by citations per capita, the United States falls roughly midway between Germany and the United Kingdom.

7 *Onderzoek Nederland*, number 408, 21 April 2017

8 Wetenschaps, Technologie en Innovatie Indicatoren, CWTS/Dialogic, 2014. The figure is taken from the original essay; hence, research disciplines are presented in the (alphabetic) order of their Dutch translations.

9 R&D-uitgaven naar sector van uitvoering en wetenschapsgebied, fact sheet Rathenau Institute, March 2017.

10 Exzellenz neu bündeln, annual oration for Max Planck Gesellschaft 2015 by Martin Stratmann.

11 *Comparative Performance of the Netherland Research Base*, Nick Fowler, Amsterdam, 25 April 2013.

12 Erasmus University Rotterdam; Maastricht University; Radboud University; University of Groningen; Delft University of Technology; Eindhoven University of Technology; Tilburg University; University of Leiden; University of Twente; University of Utrecht; University of Amsterdam; Vrije Universiteit Amsterdam; Wageningen University.

13 AMOLF Physics of Functional Complex Matter; ARCNL Advanced Research Center for Nanolithography (partnership with ASML); ASTRON Netherlands Institute for Radio Astronomy; CWI Centre for Mathematics and Computer Science; DIFFER Dutch Institute for Fundamental Energy Research; Nikhef National Institute for Subatomic Physics; NIOZ Royal Netherlands Institute of Sea Research; NSCR Netherlands Institute for the Study of Crime and Law Enforcement; SRON Netherlands Institute for Space Research.

14 Data Archiving and Networked Services (DANS); Fryske Akademy (FA); Huygens ING, research institute of History and Culture; International Institute of Social History (IISG); Royal Netherlands Institute of Southeast Asian and Caribbean Studies (KITLV); Meertens Institute for Dutch linguistics and cultural research; NIOD Institute for War, Holocaust and Genocide Studies; Netherlands Interdisciplinary Demographic Institute (NIDI); Netherlands Institute for Advanced Study in the Humanities and Social Sciences (NIAS); Hubrecht Institute for Developmental Biology and Stem Cell Research; Netherlands Institute for Neuroscience; Netherlands Institute of Ecology (NIOO); Spinoza Centre for Neuroimaging; Westerdijk Fungal Biodiversity Institute; Rathenau Institute.

15 Deltares; ECN; MARIN; NLR; TNO; WUR/DLO.

16 Royal Netherlands Meteorological Institute; National Institute for Public Health and the Environment; Netherlands Forensic Institute; Research and Documentation Centre of the Ministry of Security and Justice; Netherlands Institute for Art History; Netherlands Institute for Transport Policy Analysis; Directorate-General for Public Works and Water Management; Netherlands Environmental Assessment Agency; Netherlands Food and Consumer Product Safety Authority; Netherlands Institute for Social Research; Netherlands Bureau for Economic Policy Analysis; Statistics Netherlands; Cultural Heritage Agency.

17 The Competitive Advantage of Nations. M. Porter, Free Press, 1990.

18 The quality of life is of course also crucial, such as what the area has to offer in terms of primary and secondary education, health care, nature and culture. See, for example, Richard Florida, Cities and the Creative Class, Routledge, 2005.

19 An interesting recent study is one by the Netherlands Bureau for Economic Policy Analysis (CPB), which concluded on the basis of an analysis of patent citations that a university's impact on innovation is strongest within a radius of approximately 25 kilometres of the university. The cluster effect is no longer measurable beyond 100 kilometres from the university. See CPB discussion paper 348: Knowledge diffusion across regions and countries, evidence from patent citations.

20 Apart from mini-states like Malta, Cyprus and Luxembourg. The Netherlands has 157 kilometres of railway per 1,000 square kilometres. Source: Hoe druk is het nu werkelijk op de Nederlandse spoor? CBS, 2 March 2009.

21 Road Statistics Yearbook 2016 (p.18), European Union Road Federation.

22 Global Competitiveness Report 2015-16, World Economic Forum.

23 ACI Europe Airport Industry Connectivity Report 2017.

24 Europe's Digital Progress Report (EDPR), European Commission, 2017.

25 In June 2017 SURF started laying a 400Gbit/second network, an extremely high capacity, between Dutch educational and research institutes.

26 The fact that researchers at NWO and KNAW institutes have less liberty than researchers at institutes in other countries, and therefore also have to compete for research funding, contributes to this.

27 Nederland en de poldermodel, Maarten Prak & Jan Luiten van Zanden, Bert Bakker 2013.

28 Column in NRC Handelsblad, 3 January 2015.

29 A related concept is that of 'co-opetition'. See, for example, Brandenburger, A.M. & Nalebuff, B.J.: Co-opetition (Crown Business, 2011).

30 Research teams made up of leading scientists from various Dutch universities receive money from the government via the Gravitation programme to carry out joint multi-year research programmes.

31 For example, the Technology Foundation (STW), the Foundation for Fundamental Research on Matter (FOM) and ZonMW, which finances healthcare research. STW and FOM were absorbed into the relevant domains of the NWO in 2017; the same is planned for ZonMw in 2019.

32 Naturally, companies such as ASML and Philips make substantial investments in R&D, but they are dedicated entirely to their core business. In the former research laboratories of Bell, IBM and Philips there was a lot of room for exploratory research that was not directly concerned with the companies' core activities.

33 The importance companies attach to environments with excellent research facilities as incubators of talent and providers of an inquisitive and investigative climate also became evident during the meeting 'Een blik van buiten' organised by employers' organisation VNO-NCW (10 April 2017).

34 Another recent example is the Princess Máxima Centre for Child Oncology, which was established in association with Stichting Kika.

35 Trust: The Social Virtues and The Creation of Prosperity, Francis Fukuyama, Simon & Schuster, 1995. A similar message is conveyed in 'Why Nations Fail: The origins of power, prosperity and poverty', Acemoglu, D. & Robinson, J.A., Crown Business, 2013.

36 By comparison, the US government's Department of Energy (DOE) allocates a research budget that is one and a half times the size of the researcher-managed National Science Foundation. Hence the US government holds much sway over priorities in scientific research..

37 'Nederland hoort bij de top van de wereld', Maarten!, April 2015.

38 'What is Science?', a speech given at a meeting of physics teachers in 1966.

39 Times Higher Education World University Rankings 2016-2017.

40 The Netherlands also perform well across the board in other tables. In the QS World University Rankings 2016-2017, all thirteen universities are in the top 330, with a median ranking of 121. In the Academic Ranking of World Universities 2016 (the 'Shanghai Ranking'), twelve of the thirteen Dutch universities are in the top-400. The Open University does not take part in these international rankings.

41 The Innovational Research Incentives Scheme provides grants for individual talented researchers at various stage of their career: Veni, Vidi and Vici grants.

42 Collaboration by country, 2003-2012, as a percentage of all documents, whole counts. OECD, Compendium of Bibliometric Science Indicators 2014.

43 European Research Area Progress Report 2016, Country Snapshot The Netherlands, European Commission, DG Research and Innovation.

44 Dutch and/or English? A consideration of language choice in Dutch higher education. KNAW, July 2017.

45 Exzellenzinitiative des Bundes und der Länder (2005-2017), Deutsche Forschungsgemeinschaft, 2005.

46 The European Union's so-called 'Lisbon objectives' agreed in 2000 and aimed—initially— at 2010, later moved to 2020 because of the economic downturn.

47 Fortschritt durch Forschung und Innovation, Bericht zur Umsetzung der Hightech-Strategie. Bundesministerium für Bildung und Forschung (BMBF), March 2017.

48 www.humboldt-professur.de.

49 The Volkswagen Stiftung is Germany's largest independent organisation for the promotion of science: "impulsgebend, interdisziplinär und grenzüberschreitend".

50 Building on Success and Learning from Experience; An Independent Review of the Research Excellence Framework. An independent review of university research funding by Lord Nicholas Stern, July 2016.

51 Autumn Statement 2016. UK Treasury, November 2016. Point 16 announces two billion pounds extra for science from 2020.

52 Those with a maître de conférence teaching position give a lot of lectures; those with a CNRS position give few, if any. This distinction is far greater than in the Netherlands.

53 See, for example, the FNV analysis Werkdruk medewerkers unversiteiten ongezond. The Young Academy also raised this issue in 2017.

54 Denmark, a nation of solutions, 2012.

55 Fact sheet on government financing, Rathenau Institute, 2017.

56 The Economic Rationale for Public R&D Funding and its Impact. EU policy briefs series, March 2017.

57 Decline in funding per student, fact sheet VSNU, 2017.

58 Letter from the Minister of Education, Culture and Science to the House of Representatives, 24 May 2017.

59 PhD Students, fact sheet VSNU, 2017.

60 Matchingbehoefte bij universiteiten, fact sheet VSNU, March 2014.

61 Spinning plates—Funding streams and prioritisation in Dutch university research. Rathenau Institute, March 2016.

62 See TU Delft kan lastiger toptalent aantrekken, interview with Tim van der Hagen, ScienceGuide, 31 August 2016.

63 We generally avoid the terms 'free' and 'unbound' research because all research in the Netherlands is in reality embedded in wider networks. Furthermore, researchers who are inspired by specific applications also have the freedom to follow their scientific nose.

64 Naar de top - de hoofdlijnen van het nieuwe bedrijfslevenbeleid, letter from the Minister of Economic Affairs, 6 February 2011.

65 Aanvraagdruk bij NWO, Rathenau Institute fact sheet, March 2017.

66 Researchers in the Life Sciences in particular also face a lot of bureaucracy: the rules for research with test humans (Central Committee on Research involving Human Subjects), for research with test animals (Central Animal Testing Committee), rules relating to genetic modification, privacy, data policies, etc. With this observation we are not saying that the rules are unnecessary, but that they do make a substantial contribution to the pressure on researchers.

67 A recent evaluation of TO2 institutes concluded that because of financial pressure there was insufficient room to conduct social research for which there is no direct demand (and therefore no funding) from business (conclusion 7).

68 At UMC Utrecht for example, see 'Weg met die publicatiedwang, zegt UMC Utrecht', NRC Handelsblad, 26 October 2016.

69 'Het prof. dr. Dagobert Duck-effect; Veelverdieners in de Nederlandse wetenschap'. Klaas Landsman in de Volkskrant of 21 February 2015.

70 Koersvast, Aanbevelingen ter verdere versterking van de bètadisciplines natuur- en scheikunde. Report of the Breimer Committee, January 2016.

71 TU Delft kan lastiger toptalent aantrekken. ScienceGuide, 31 August 2016.

72 Beloning van wetenschappelijk personeel in internationaal perspectief. SEO Economisch Onderzoek, September 2015.

73 "Je ziet jonge wetenschappers afhaken. Die zeggen: ik wil deze rat race niet meer, deze struggle voor geld." Interview with Bert Weckhuysen, Financieele Dagblad, 27 August 2016.

74 De aantrekkelijkheid van Nederland als Onderzoeksland, KNAW, due in autumn 2017.

75 'De beste beleidsbuffer is een apart fonds for infrastructuur, onderwijs en onderzoek'. Financieele Dagblad, 15 April 2017.

76 Investment Agenda, Dutch Knowledge Coalition, 15 September 2016.

77 The Knowledge Coalition is made up of the universities (VSNU), the universities of applied sciences (Vereniging Hogescholen), university medical centres (NFU), KNAW, NWO, VNO-NCW, MKB-Nederland and the institutes of applied research (TNO/TO2).

78 See, for example, the publication 'Rigor Mortis' by Richard Harris: http://richardharriswrites.com/

79 OECD Main Science and Technology Indicators, 2016

80 Since recently the Netherlands does have Ammodo, the institute of arts and sciences, which offers awards to stimulate fundamental scientific research. The Gieskes Strijbis Fonds also subsidises breakthrough projects. The VSBfonds, which used to finance scientific projects, now focuses on social and cultural projects.

81 Building on the final report of the task force Geven for Weten, private funding for science.

82 The societal impact of collaboration, and hence the potential of the NWA, is further underscored by the findings and recommendations in the recent position paper *Productive interactions: societal impact of academic research in the Knowledge Society* from the League of European Research Universities (LERU), March 2017.

83 Many general universities in the US, such as Harvard, have a school of engineering and applied sciences. In the Netherlands, the University of Groningen has an engineering programme.

84 Chemistry & Physics, Fundamental for our Future, vision document by the Dijkgraaf Committee, 2013.

85 The evaluation of the TO2 institutes in April 2017 also stressed the importance of cooperation throughout the chain: "Cooperation between TO2 institutions and other knowledge players is essential for the future" (recommendation 14).

86 Examples are Danone (Utrecht), FrieslandCampina (Wageningen), the consortium of Microsoft, QuTech and FOM/NWO (Delft), ARCNL (a consortium of ASML, VU Amsterdam, University of Amsterdam and FOM/NWO in Amsterdam) and the ARC Chemical Building Blocks Consortium with hubs in Groningen, Eindhoven and Utrecht and supported by AkzoNobel, BASF, and Shell.

87 Science Park gaat in zaken, Financieele Dagblad, 30 March 2017.

88 Such suggestions are also not financially feasible. In 2016, Harvard controlled endowments worth 35.7 billion US dollar.

89 Ruimte voor ongebonden onderzoek, KNAW advisory report, 2015.

90 By analogy with the Industrial Partnership Programme of NWO/FOM.

91 Vision for Science 2025: choices for the future, the government's reaction to the Interdepartmental Policy Review on Scientific Research (2014) and the reports of the Advisory Council on Science and Technology Policy (AWTI) entitled 'Standing out from the crowd' ('Boven het maaiveld', 2014) and 'Size and suitability' ('Maatwerk in onderzoeksinfrastructuur', 2013).

92 See, for example, the University of Groningen's Rosalind Franklin Fellowships, a programme that has been replicated by various universities and the NWO institutes.

93 Fundamentele wetenschap is in verdomhoekje beland, interview with Marc Timmers, Financieele Dagblad, 16 February 2017.

94 Chemistry and Physics: fundamental for our future, Dijkgraaf Committee, 2013. Dijkgraaf also made the suggestion at the VNO-NCW meeting 'Een blik van buiten' in April 2017. See also WTI-Diplomatie – Offensief voor internationalisering van wetenschap, technologie en innovatie, AWTI, May 2017.

95 See: http://energy.mit.edu

96 Topfysica samen met bedrijven; Evaluatie van het FOM Industrial Partnership Programme, April 2015.

97 Trust in Science ('Vertrouwen in wetenschap'), KNAW advisory report, 2015.

Photo credits

All photos were acquired from istockphoto.com, except page 72, which was taken from 'Onderwijs-prijs KNAW, Room for ID's'.